Nebulae

Illustrations: Janet Moneymaker
Design/Editing: Marjie Bassler

Nebulae
ISBN 978-1-953542-23-6

Published by Gravitas Publications Inc.
Imprint: Real Science-4-Kids
www.gravitaspublications.com
www.realscience4kids.com

Image credits
Cover: Cat's Eye Nebula, Courtesy of Hubble/NASA, ESA, HEIC, and The Hubble Heritage Team (STScI/AURA)
Above: Pillars of Creation: NASA and ESA, Hubble Space Telescope (STScI)

Today, **astronomers** use **space telescopes** to take pictures of planets, stars, comets, and asteroids.

We use telescopes on Earth too.

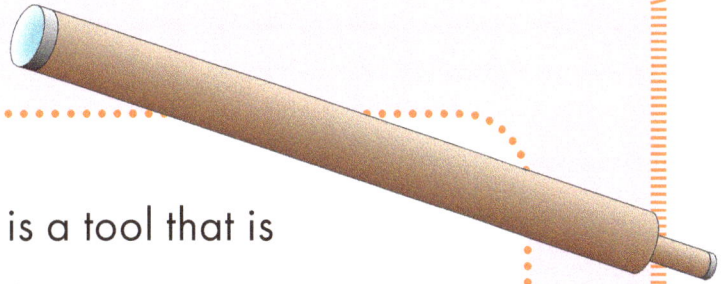

A **telescope** is a tool that is used to make faraway objects in space look bigger.

A **space telescope** is placed high above the surface of Earth. It travels around Earth and gathers information about objects in space. This information is sent back to astronomers on Earth.

The **Hubble Space Telescope** has been in use for many years. It has taken over a million pictures of objects in space. Some of these are pictures of **nebulae.**

What do you call more than one **nebula**?

Nebulae!

Say... (NE-byuh-luh) (NE-byuh-lee)

A **nebula** is a huge cloud of gas and dust.

Beautiful!

The gas and dust swirl together
to create beautiful cosmic art.

Some nebulae are areas in space where new stars are formed.

I never thought about where stars come from.

The blue dots are young, very hot stars.

Star Forming Area: Hubble/NASA, ESA, F. Paresce (INAF-!ASF, Italy), the WFC3 Science Oversight Committee

Some nebulae are formed when a dying star explodes and turns into gas and dust.

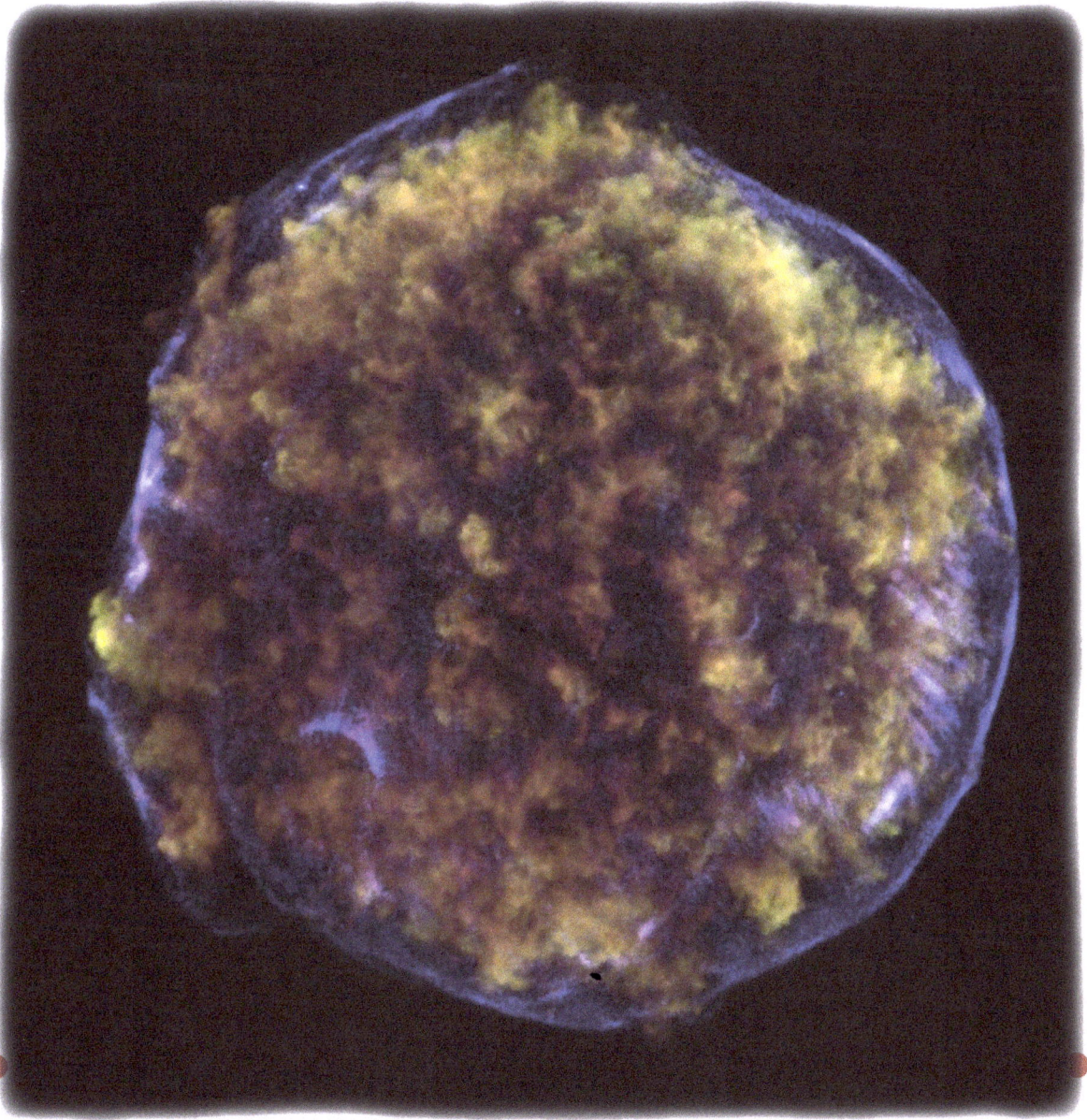

Tycho's Supernova Remnant: NASA/CXC/Chinese Academy of Science/F. Lu, et al.

The **Helix Nebula** is the closest known nebula to Earth.

Even the closest nebula is very, very far away.

Helix Nebula: Hubble/NASA, ESA, C.R. O'Dell (Vanderbilt University), The Hubble Heritage Team (STScI/AURA

The **Cat's Eye Nebula**

is famed for its beauty.

Cat's Eye Nebula: Hubble/NASA, ESA, HEIC, and The Hubble Heritage Team (STScI/AURA)

The **Eagle Nebula** has an interesting part called the **Pillars of Creation.**

Amazing!

Pillars of Creation: NASA and ESA, Hubble Space Telescope (STScI)

Nebulae are fascinating
to observe and study.

Is that called the Bad Hair Day Nebula?

I do not think so.

Supernova Remnant N49: NASA and The Hubble Heritage Team (STScI/AURA)

How to say science words

asteroid (AA-stuh-royd)

astronomer (uh-STRAH-nuh-mer)

comet (KAH-muht)

creation (kree-AY-shuhn)

eagle (EE-guhl)

galaxies (GAA-luhk-seez)

Helix (HEE-liks)

Hubble (HUH-buhl)

nebula (NE-byuh-luh)

nebulae (NE-byuh-lee)

pillar (PI-luhr)

planet (PLAA-nuht)

science (SIY-uhns)

space (SPAYSS)

star (STAHR)

telescope (TEL-uh-skohp)

universe (YOO-nuh-vuhrss)

What questions do you have about NEBULAE?

Learn More Real Science!

Complete science curricula from Real Science-4-Kids

Focus On Series

Unit study for elementary and middle school levels

Chemistry
Biology
Physics
Geology
Astronomy

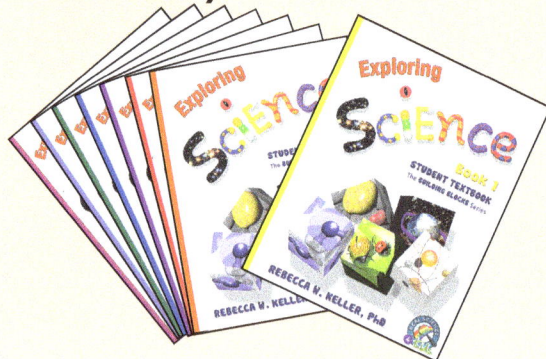

Exploring Science Series

Graded series for levels K–8. Each book contains 4 chapters of:

Chemistry
Biology
Physics
Geology
Astronomy